Fancy Fences & Ga...
Great Ideas for Backyard Carpenters

Tina Skinner

Schiffer Publishing Ltd

4880 Lower Valley Road, Atglen, PA 19310 USA

Acknowledgments

All of the images and graphics in this book were provided by the California Redwood Association, as well as the tips, graphics, and instructions for building and installing a fence or gate.

Additional assistance was provided by the American Fence Association.

Library of Congress Cataloging-in-Publication Data

Skinner, Tina.
Fancy fences & gates : great ideas for backyard carpenters / Tina Skinner.
   p. cm.
ISBN 0-7643-1417-3 (Paperback)
1. Wooden fences--Design and construction--Amateurs' manuals. 2. Gates--Design and construction--Amateurs' manuals. 3. Carpentry--Amateurs' manuals. I. Title.
TH4965 .S55 2001
631.2'7--dc21
2001001547

Copyright © 2001 by Schiffer Publishing, Ltd.

All rights reserved. No part of this work may be reproduced or used in any form or by any means—graphic, electronic, or mechanical, including photocopying or information storage and retrieval systems—without written permission from the copyright holder.
"Schiffer," "Schiffer Publishing Ltd. & Design," and the "Design of pen and ink well" are registered trademarks of Schiffer Publishing Ltd.

Designed by Bonnie M. Hensley
Cover design by Bruce M. Waters
Type set in Aurora BdCn BT/Zurich BT

ISBN: 0-7643-1417-3
Printed in China

Published by Schiffer Publishing Ltd.
4880 Lower Valley Road
Atglen, PA 19310
Phone: (610) 593-1777; Fax: (610) 593-2002
E-mail: Schifferbk@aol.com
Please visit our web site catalog at **www.schifferbooks.com**
We are always looking for people to write books on new and related subjects. If you have an idea for a book, please contact us at the above address.

This book may be purchased from the publisher.
Include $3.95 for shipping.
Please try your bookstore first.
You may write for a free catalog.

In Europe, Schiffer books are distributed by
Bushwood Books
6 Marksbury Ave.
Kew Gardens
Surrey TW9 4JF England
Phone: 44 (0)208 392-8585
Fax: 44 (0)208 392-9876
E-mail: Bushwd@aol.com
Free postage in the UK. Europe: air mail at cost.

Contents

Introduction	4
Tips for Designing and Building	5
A Gallery of Gorgeous Fences and Gates	18
Resources	80

Introduction

The proverbial "perfect home" is rarely described. Instead we simply speak of a "house with a white picket fence." That pretty picket fence image is how we visualize contentment, a well-contained abode peacefully set apart for our pleasure.

A fence is one of the key features of our outdoor environment. It sets the boundaries of what is ours vs. off limits. And it provides a first impression to any onlooker. So it's rather amazing that there are so few resources available for anyone looking for ideas for fences. This book is meant to be just that – a treasure trove of ideas for someone looking for the perfect perimeter for their backyard. These gorgeous fences and gates will encourage you to get to work creating your own unique fence or gate. Moreover, useful tips and instructions will help you plan and execute the building of your own fence and gate.

Tips for Designing and Building

Designing your fence

This book is designed as a visual guide to help you choose the style of fencing most appropriate for your property, your tastes, and your needs. A fence is an extension of your home and calls for as much attention to its purpose, style, and design as you would give if you were adding another room.

To get started, you should ask specific questions: Why are you building the fence? Are you defining a boundary? Are you looking to create more privacy or to shield yourself from unwanted noise? Do you want to add a windbreak, create shade for a deck, or contain the family dog?

Next you'll want to ask yourself how the fence will complement the character and design of your home. Consider how your neighbors will view the fence. Some fences look wonderful on your side, yet present neighbors with a less appealing view. Traditionally, the best side of a fence is faced out, toward the neighbors. One way to start might be by chatting with your neighbors. A friendly discussion might even lead to all parties contributing to the cost of materials and labor.

There are a surprising number of fence styles to choose from, many of which are included in this book. You may choose an existing design or you can create your own from scratch. You can also modify a basic design into a look that is distinctively yours. You'll find it fun to experiment; even small variations will make a big difference. Keep in mind that you only need to plan the details for one typical bay (the section of fence from one post to the next). From there you can calculate everything else you need for a fence of any length.

If you are working within a limited budget, you can still build a fence that does the job without sacrificing quality by choosing a style that uses less lumber or a more economical grade of lumber. You can also bring down costs by making the best use of standard lumber lengths.

Establish your layout priorities

A fence can impact your site in many different ways, so in your planning you need to think about which aspects of the site you want to retain and which you'd like to change. Before you finalize your fence line, carefully review your priorities to confirm that all your key considerations are being covered. Some of the questions you'll want to answer are:

· Which areas do you want to keep or block?

· Where are all the activity areas and what are the traffic patterns in those areas?

· What needs to be protected from the sun or prevailing winds?

· Is there noise you'd like to block? Where is it coming from?

· Do you want the fence to support certain vines or shrubs?

· Are there places where the fence must be in scale with existing landscaping, or coordinated with existing structures like trellises and planters?

Draw a site plan

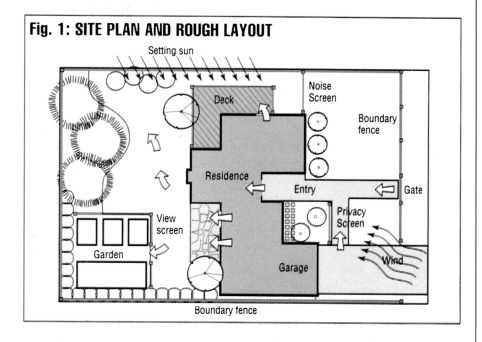

Fig. 1: SITE PLAN AND ROUGH LAYOUT

It is easier and less costly to resolve all your key issues while you're still at the conceptual stage. To do this, you need to have a site plan. You can draw a site plan in an hour or two, or you can look for a copy of an existing plan with the building department, designer or architect, building contractor, landscape contractor, or even a previous owner. Once you have a plan of your property, you can experiment with different fence line schemes.

Start by making a rough sketch of the site, including such things as site perimeter, the house plan, driveway, patios, walkways, garden beds, and utilities. You will also want to indicate grade, sun and wind orientation, and the characteristics of surrounding structures and plantings. In plotting out the sun's path, visit the site several times during the day to note how shadows fall. And don't forget that the sun follows a different path during different seasons.

Use circles for activity areas, arrows for traffic paths, wavy lines for winds, a yellow dotted line for the sun's path, squiggly arrows for noise direction, different color arrows for good and poor views, or whatever other symbols you find most practical. Now walk the property with a 50- or 100-foot tape and someone to hold the other end. Precision counts. Take actual field measurements and note them on your sketch.

Once you're done, transfer your field measurements to 1/4-inch scale graph paper. Use light pencil lines initially to plan your fence line and only darken them when everything is worked out to your satisfaction. Also, it's a good idea to jot down the dimension next to each line. Later on it will be easier to read than counting squares.

Make a rough layout

Once you have everything indicated, it's time to start looking at fence lines. Since you're just playing with ideas at this point, don't overwork any one layout scheme. Use tissue overlays to make various rough fencing schemes.

A good rule of thumb is to keep spaces as large as possible. Most people enjoy the expansiveness of the great outdoors, so think twice before you plot a fence line that boxes people in.

If you have gates, don't forget to allow enough room for openings. Three foot openings are generally enough to accommodate one person or a large piece of equipment. Four feet will accommodate two people while 6 feet is generally sufficient for a group. For vehicles you want to allow a minimum of 10 feet.

Once you're satisfied that you have what you want, trace your rough layout and site plan on another tissue overly. Indicate where the openings will be and if there's a gate, the direction in which it will swing. Completing this step will help you calculate how much lumber will be needed to build your fence.

If you see ways to improve your fence line, don't hesitate to make new sketches until you feel you've gotten everything just right.

Legal considerations

It's a good practice to check out the legal considerations associated with building a fence. Local codes and ordinances can vary considerably from one community to the next. Most communities have height restrictions on boundary or division fencing. There also may be certain laws and codes that actually require you to erect a fence, for instance, around a swimming pool or open well.

If there are any questions about whose land the fence is being built on, arrange for a survey. If any part of the fence encroaches on your neighbor's property, you may be asked to move it.

Fig. 2: LUMBER GRADES AND CHARACTERISTICS

		Heartwood	Sapwood
Architectural Grades	Clear Limited Knots	Clear All Heart B Heart	Clear B Grade
Garden Grades	Knotty	Construction Heart Deck Heart Merchantable Heart	Construction Common Deck Common Merchantable

Building your fence

Fence building is divided into three stages: First you physically plot the fence by staking out the location of the posts, then you install the posts, and finally you add the rails and fence boards.

Most people prefer to set all the posts in place and then attach the rails and fencing, especially when the posts are being set in concrete. Another approach is to assemble the rails and fence boards whenever two posts are in place. This has certain advantages when you're working with prefabricated fences, or when you first build each section on the ground and then lift it into place.

For the first step, locate the exact course your fence will take and mark the line with stakes and string. This is the most exacting part of the project because it establishes the foundation and framework for your fence.

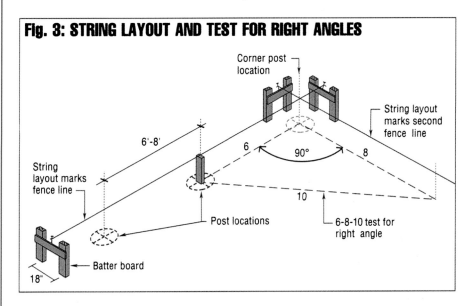

Fig. 3: STRING LAYOUT AND TEST FOR RIGHT ANGLES

To plot a straight line for your fence, mark the location for each end or corner post with a batter board (two solidly driven stakes 18 inches apart and connected by a 1x3). Use the center of the batter board as your point of alignment and drive a nail or cut a notch. Tie a piece of mason's twine or string to the nail of one batter board, draw it taut, and tie it to the nail on the other. If the fence line is particularly long, you'll want to support the twine with stakes whenever the twine begins to sag.

Posts are generally spaced 6 to 8 feet apart, depending on the style of fence. Measure and mark the center locations of all the posts with chalk or a pen. Corner posts will be located directly under the crossing string lines at the batter boards. Take the string of your plumb line, line it up directly with the first chalk marker. Mark where the point of the plumb bob falls, using a spot of spray paint or a stake stuck into the ground. Paper nailed into the ground can also serve as a marker. Once the center marks for all the posts are indicated, you can untie the string.

If your fence includes 90-degree right angles, they can be accurately determined by using the 6-8-10 triangle measuring technique shown on page 9. Any multiple of 3-4-5 will work, with larger numbers being easier to measure.

Establish the first fence line as explained above, then establish a second fence line roughly perpendicular to the first using another batter board. Measuring from the stake that will form the corner, place a chalk mark 6 feet away along the twine that forms the first fence line. Next, put a mark eight feet away from the corner stake on the twine forming the second fence line. Finally, measure the distance between the two chalk marks and adjust the second fence line on the batter board until the diagonal measurement between the two marks equals 10 feet. This gives you an accurate 90-degree angle.

Set posts

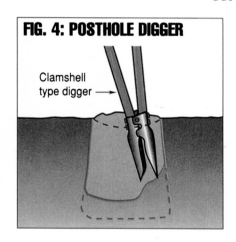

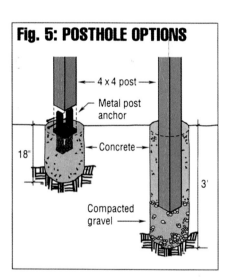

Now comes the hardest part of building a fence – digging the holes and setting the posts. For the first step, a post-hole digger is often all you need. However, if there are many holes to dig, you may want to consider one-man or two-man power augers. Augur-type diggers are good for rock-free earth, but if you're likely to encounter large stones, a clamshell type is better. A digging bar is also useful for prying rocks and other hard-to-move debris from the hole.

In most cases, your posts will be 4x4s or 6x6s, depending on fence style. Posts can be set directly into the concrete or attached with metal post anchors. Posthole diam-

Build and hang the gate

There's something deeply satisfying about a gate that opens easily, swings freely, and closes securely with a reassuring click. Building such a gate requires you to exercise care and craftsmanship in each of the five steps of gate construction: setting the gate posts, building the frame, adding fencing boards, hanging the gate, and installing the latch.

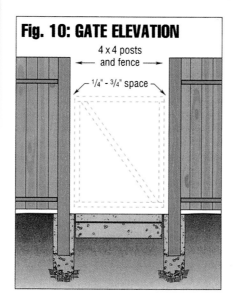

Fig. 10: GATE ELEVATION

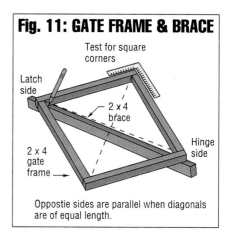

Fig. 11: GATE FRAME & BRACE

Opposite sides are parallel when diagonals are of equal length.

Gate posts should be set deeper than your fence line posts – about 1/3 their total length, and anchored in concrete. This is necessary because of the need to withstand additional stresses. Posts also must be carefully plumbed so that the inside faces are exactly parallel.

When measuring the opening, make sure to allow for clearances on the hinge and latch side of the gates. For gates with standard 2x4 framing and 4x4 posts, leave a 1/2- to 3/4-inch space between the latch post and the gate frame. On the hinge side, it will depend on the hardware you use. About 1/4 inch is usually sufficient.

Once the posts are set, begin assembling the frame. Cut the pieces to length and assemble them on a flat surface, making certain that the gate frame is in square. Use a carpenter's square to check the corners and then measure the diagonals. When the diagonals are equal, the sides of your gate are parallel and the frame forms a true rectangle. Use wood screws and good exterior wood glue instead of nails for added strength.

To measure the brace, lay the frame down on top of the 2x4 bracing member and mark your cut lines. The easiest cut is a single, angled cut so that the brace will run from hinge side bottom to latch side top. Cut the 2x4 just outside your marks so the brace will have a tight fit, and attach the brace to the frame with nails or screws.

Now add the boards, starting from the side where the hinge will go. If the last piece is not flush with the frame edge, either space the boards slightly or plane a little from each board until they fit. Then drill your pilot holes and fasten the hinges to the gate. There are a large variety of hinges and latches to choose from. The graphic below shows the most common.

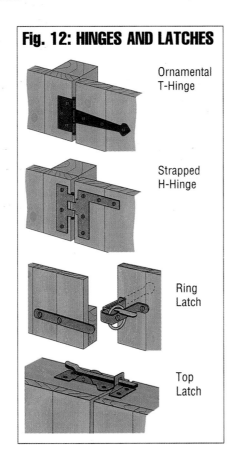

Fig. 12: HINGES AND LATCHES

Ornamental T-Hinge

Strapped H-Hinge

Ring Latch

Top Latch

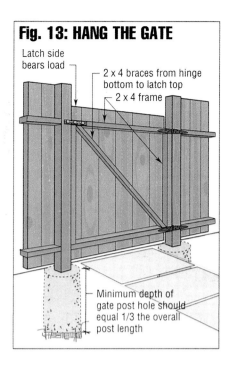

Fig. 13: HANG THE GATE

At this point you are ready to hang the gate, but before you do anything, you'll want to check the fit by moving the gate into position and trimming any areas that are too tight to provide ample clearance. Now prop the fitted gate into the opening using blocks to support it, or have a helper hold it in position, and mark the hinge and screw hole positions on the post. Once the holes are drilled, replace the gate, and attach the hinges to the post. Finally, mount the latch assembly on the gate and post, using screws a little longer than usual to help the latch withstand the punishment it will take through years of use.

Finally, it's time to congratulate yourself on a job well done.

A Gallery of Gorgeous

Board sizes were varied in this privacy fence to create panels, a subtle design effect that deflects rather than draws attention.

Fences and Gates

Plastic sheathing imitates rice paper for a look straight out of Japan.

This staggered wall of substantial redwood boards defines a property line, and creates a view. It is both gateway, planter, and sculpture, designed by Landscape Architect Richard Splenda.

A stretch of privacy fence is punctuated with "window boxes" for hanging or stationary plant displays. The inserts create an architectural element for the yard.

The diagonal is emphasized in panels for this handsome privacy fence. Posts, cap, and skirt-board are 4x6s and the diagonal 1x4 boards are held between frames of 2x4s The corner brace hardware functions only as a design element.

A colorful effect was created by marrying different grades of redwood for fencing and deck here. Each slat in the tall privacy fence was made from two different-size boards, adding texture to the "wall.

An imposing, rustic, rough-textured fence helps screen a front entryway.

25

A series of four-foot-high sections step down a sloping property line. Design interest was added with a round-top gate bordered by an antique pillar.

An intricate cutout pattern allows light through to come through this privacy screen, and evokes the Orient.

This privacy fence is topped with appropriately substantial lattice work.

This fence creates an outdoor room for the owners of a double Victorian in San Francisco. Detailing in the lattice work and post caps capture the Victorian flavor.

A little is all you need in a lot! Here an outdoor environment is made very inviting with big flagstones, and intimate table, and a privacy fence. Board sizes in the six-foot-high fence alternate from 4-inch, 6-inch, and 8-inch planks. A handsome redwood gate topped by a trellis secures the area.

A slatted fence imitates the mountainous horizon, punctuating this desert landscape with rich redwood hues. The gate opening and top of the fence were finished with 1/2x4 redwood bender board, which also lines the curved walkway.

New fencing incorporates trellises, benches, and planters — all constructed in matching redwood to create a coordinated design around this home.

A simple turn-of-the-century white cottage is framed by artful landscaping, including latticework fences, gates, and an arbor entry.

The most successful fence and gate solutions create a sense of arrival and anticipation. This inviting entryway features a lacy latticed and curved gate topped by an imposing pergola. A matching fence was built 6-1/2 feet high to provide privacy and security.

Preexisting brick pillars were topped with wood column boxes and capitals and connected by a fence designed in the Arts and Crafts style.

Cement columns were capped with copper lighting units and framed by handsome redwood gate and privacy fence panels.

These homeowners wanted a structure that would define a section of their garden and frame a pleasant tree-scape. Custom craftsman Julian Hodges created this fence with a scoop-design gate and overhead trellis.

Bamboo inserts accentuate the Asian influence in this straightforward fence and gate. The ends of the sturdy overhead beam are trimmed in traditional Craftsman "cloud climb" details to lighten the design.

A six-foot-wide pedestrian entry gate opens to a lush front lawn and a wide walkway beneath a soaring trellis. Hand-carved quatrefoil detailing – a classic Gothic tracery design – adds whimsy as well as practical peep holes. The arched beams on top of the gate were cut from 3x10s into the 2x5 arched shape.

Rustic knots and streaks of cream-colored sapwood characterize economical garden grades of redwood. These color variations are an inherent part of the design of a privacy fence and deck railing.

This classic redwood Victorian fence was built to duplicate the original fence erected in 1881, working from a historical photograph and a few existing pickets. A 12-inch grass board was attached to the outside of the posts and the bottom rail to-nailed into place above it.

An elegant yet simple post-and-rail fence echoes design elements of the Craftsman style and of the Japanese garden it frames. Designer Timothy A. Jones added extra interest was added by alternating the sizes of the fence boards, which were spaced for an airy feeling.

A series of fence panel modules are stepped along a sloping site. Design interest was added to the top of the fence with alternating inserts of upright boards placed sideways and whimsical gabled roofs.

A gate and fence create an impressive pedestrian garden entrance. Alternating board widths, fence eight, and lattice panels create variations in this design by Timothy Jones of Calasian Hardscapes.

A massive rooftop-style canopy shelters an entry gate, built in the Craftsman style by Timothy Jones.

Neighborhood pets were not welcome in this yard, so craftsman Timothy Jones created an unobtrusive design to keep all but the best jumpers out. The four-foot-tall fence and driveway gate were topped with a concave rail.

This Craftsman-style fence and gate is both beautiful and functional. The solid board fence and gate are softened and balanced by a redwood trellis supporting wisteria vines. A low companion screen conceals garbage cans.

This 10-foot tall fence, gate, and trellis accomplishes two goals: creating both a windbreak and a privacy screen for a backyard patio. Copper sheets were mellowed with chemicals to create a blue/green patina and add interest to the structure in the top panels as well as inserts in the gate accent.

The diagonal is emphasized in both latticework and a privacy screen, with alternating vertical plank walls. The fence is part of a patio entryway combining seating, planter boxes, and a trellis/grape arbor for a true feeling of arrival.

An inviting, private setting was created with a louvered, seven-foot redwood fence. A top rail adds interest.

Pretty enough to be paneling inside the house, this three-tiered fence creates a visual treat outside. Tongue-and-groove boards are topped by a strip of lattice and then by louvers set at a 45-degree angle, enabling the owners to see out, but preventing passersby along the street from seeing into the yard. Inviting niches for hanging plants were created inside the columns.

Opposite pag: A built-in bench and latticework shade shelter add style and interest to a long stretch of privacy fence.

A rustic yet contemporary six-foot-tall grapestake fence creates security and privacy, enhancing an English-style cottage and setting off brick columns.

A pleasing blend of natural colors and textures is achieved by combining redwood, copper aged to a blue-green patina, and brick. The six-foot security fence provides privacy for a front yard as well as a buffer for street noises on a busy main artery.

This massive, stockade-like design is a redwood board-on-board "good neighbor" style, which looks good from either side.

This otherwise ordinary six-foot-tall fence got added eye appeal when alternate boards were cut in the dog-ear pattern.

Design wasn't sacrificed when this fence was installed for both privacy and security: the owners invested in a beautiful cast iron gate and lamp, and adorned the fence with latticework and a trellis.

Opposite page: The 1x4 fence boards were angled and overlapped slightly in this louvered privacy screen to allow filtration of light and air.

Opposite page: Intricate screens provide privacy and visual interest and complement overhead trellises, built-in planters, and benches in this deck environment.

When planning a new privacy fence, the owners of a stylish Tudor house in Berkeley, California, asked designer Marty Reutinger to incorporate some cast iron Art Deco rattlesnake sculptures they had acquired years earlier. The result is a six-foot-tall fence that flows in six-foot sections. Irregular "grape stakes" were used for fence boards to add rustic appeal, and grillwork panels frame the iron sculptures, which were painted a faux verdigris. An arched redwood trellis over the recessed gateway provides a visual focus point.

An artfully curved gate top and imposing shaped post caps characterize this elegant entryway.

Opposite page: An imposing, oversize trellis arcs over this patio entryway, topping an artful slatted fence and gate with varying board widths on the vertical, arches and straightaways on the horizontal.

The design of this fence was inspired by a photo study of fences and gates in Greece. Scott Smith wanted the overall structure to be human in scale and to suggest permanence without creating the feeling of a barricade. He repeated an eight-foot module along the 140-foot length of the redwood fence, borrowing simple geometric shapes from his Craftsman-style bungalow.

An arbor serves as transition from the front entry to the backyard garden, in a series of six arches with a gate.

An outdoor spa and deck "room" is surrounded by a privacy fence designed to allow light and air filtration. The boards maintain their horizontal alignment up a steep incline.

A fence and gate create a transition between pool area and a Japanese garden and koi pond. The interrupted motif of the railings is the Japanese symbol for clouds. Cedar shingles top the Shinto-style entryway and a blonde bamboo insert creates an appealing skirt, contrasting with the rich tones of redwood.

A swirling, curved abstract sculpture is created in a gate. Oval and crescent cutouts near the arched top form a focal point

Opposite page: An oval cutout and rounded edges soften and add style to a beautiful gate. The gate is attached to a simple yet elegant "good neighbor" style redwood board-on-board fence decorated with routed-out detailing on the posts and fence tops.

Traditional New Orleans detailing was used for this imposing double gate. The intricately crafted raised panel gate was built using mortise and tenon construction techniques.

Because this gate gives one their first impression of the home, architectural woodwork specialist Julian Hodges designed a redwood gate and fence reminiscent of traditional Japanese structures. The structure offers privacy and security while being inviting at the same time.

Resources

The California Redwood Association can be contacted with regard to the designers and builders of the fences and gates pictured in this book. Also, the American Fence Association, Inc., can provide information about fence manufacturers and installers in your area.

California Redwood Association
405 Enfrente Drive, Suite 200
Novato, CA 94949
415-382-8531
www.calredwood.org

American Fence Association, Inc.
2336 Wisteria Drive, Suite 230
Snellville, GA 30078
800-822-4342
www.americanfenceassociation.com